Anonymous

Astounding Disclosures!

: Hell on earth! Murder, rape, robbery, swindling, and forgery covertly

organized! Cannibalism made dainty! An exposition of the infernal

machinations and horrible atrocities of whited sepulcherism

.

Anonymous

Astounding Disclosures!
: Hell on earth! Murder, rape, robbery, swindling, and forgery covertly organized!
Cannibalism made dainty! An exposition of the infernal machinations and horrible
atrocities of whited sepulcherism

ISBN/EAN: 9783337164102

Printed in Europe, USA, Canada, Australia, Japan

Cover: Foto ©berggeist007 / pixelio.de

More available books at **www.hansebooks.com**

'UNDING DISCLOSURES!!

243

:th! Murder, Rape, Robbery, Swin-
..nd Forgery Covertly Organized!
Cannibalism Made Dainty!

AN EXPOSITION

OF THE

\IFERNAL MACHINATIONS

AND

HORRIBLE ATROCITIES

OF

...ITED SEPULCHERISM;

TOGETHER WITH

A SURE PLAN

FOR ITS

SPEEDY OVERTHROW.

BY THE AUTHOR OF

"THE RELIGION OF SCIENCE," "THE ESSENCE OF SCIENCE," AND
"THE NEW CRISIS."

icility

NEW YORK:

..ISHED BY CALVIN BLANCHARD

76 NASSAU STREET.

1860.

ARGUMENT.

I. Supernatural foolery, the foundation of legal and political humbug; whence result human degradation and misery.

II. The cunning hypocrisy and treachery of theologians, and the covert scoundrelism of law-mongers and politicians, made plain to all victims who dare think.

III. Truth, reason and goodness, *unorganized*, can never successfully oppose firmly cemented and well organized foolery, humbug and evil.

IV. How science, as an organized whole, and installed as both religion and government, can redeem mankind, and establish freedom and happiness.

HELL ON EARTH.

THE immediate importance of my subject admits of no preliminary.

Having ascertained when and where the "Whited Sepulcherites" held their annual conclave, and knowing, furthermore, that the priests of Woollywolf* had free access to all the meetings of this infernal brotherhood, I furnished myself with the sacerdotal costume, and boldly ventured into their

PANDEMONIUM.

I found the company composed of about an equal number of the most respectable looking and elegantly dressed gentlemen and *ladies!*

In justice to the ladies, I must say that they were not *active*, but only *passive* members of the society.

The President rose, and said—

Ladies and Gentlemen; and (*Bowing low*,) Most Reverend Priests—Never yet have I appeared before you so completely entranced in the most exquisitely infernal delight.

Our aims are all accomplished. Our purposes all fulfilled. Our work completed.

The nineteenth century pays roundly for, and gratefully wraps itself in, the tattered swaddling clothes of society's mental infancy, and tremblingly bows to the earliest, and therefore most false and barbarous opinions ot which but the veriest shreds have been preserved. Consequently—

Charlatanry and imposture, backed by blind prejudice. and indirectly but most effectively assisted by short-sighted negativism, keep science the headless and therefore all but useless thing, it must remain 'til ex-

---

* The name of the chief deity of the oldest, and consequently most barbarous religion of which any knowledge has descended to our time. This Theological fossil, is the base of Whited Sepulcherism.

tended into the political, moral and social world.　Where, now, in its stead,—

Confusion and anarchy are enthroned as law and order.

Folly triumphs over wisdom.

Ignorance lords it over knowledge.

Wrong, magnificently great, has usurped the place and even the name of right, and imposed its responsibilities and odium on mere peccadillo.

Liberty consists but in man's right to tax himself for his own enslavement, and for converting the world into a " vale of tears " through which he bustles in less than one fourth the time he would, guided by science, remain supremely happy therein.

Even love is so nearly dethroned that lawyers, constables and turnkeys are the graces which minister at the court of Cupid, whose wings are cut off, whose eyes are put out, and whose heart is tightly wrapped in crusty vellum, or firmly bound in law sheep.

Free-lover (as if there could be any other kind of lover) is an opprobrious epithet among the beautiful half of humanity.

Free-thinker (as if there could be any other kind of thinker) is an opprobrious epithet among men.

Mankind have left them but the dismal alternative of splendid misery or squalid wretchedness.

What can concentrated evil more ?

What more would, even Whited Sepulcherism ?

Infernal joy thrills my very marrow.

[*Demonstrations of the wildest delight.*]

What, compared to ours, are the stale joys of the worshippers of free discussion ?

At the first glance, one would be tempted to wish that Scratchfyre * and his regions were a reality, for the sake of these stubborn oppositionists. [*Here the young men set up such a tiger as would have done credit to all the wild beasts in Africa, and the ladies*

* The name of their chief devil, sometimes nicknamed Old Scratch.

looked as though they wished they were of the harsher-voiced sex.]

Nay, after all, these unconnected zealots are hardly worth our spite. True, their numbers are very respectable, and their *individual* wealth still more so. But what need regular troops, with able leaders, care for unorganized masses, however numerous? The property of infidels is taxed, *sub rosa*, in a hundred different ways, for our support. Were they ten times as wealthy as ourselves, it would avail them nothing. Nay, it would strengthen our *organization* more than it would their *opposition* to it, so long as the watchful priests of Woollywolf, whose interests are identical with ours, have the shaping of the human organs of thought, and, *sub rosa*, saturate with their doctrines, all that is taught in the public schools, for the support of which all are taxed, and which taxes our negativists pay without *much* grumbling. Our trusty priests, I say, can, and will, so form the infant mind, or rather so deform it, [*Laughter*] that the multitude can no more reason or understand than fashionable Chinese ladies can race.

We must not, however, conceal from ourselves the fact that in these schools, are forged weapons which would prove dangerous to us if used, But the mere means of acquiring the useful, we represent as the useful itself, and there the matter thus far ends.

But the funniest thing of all, is, that these opponents of ours, who are so boastful and vain of their superior reason, support, as heartily as do the average of our adherents, one of the most effective of all our measures, the object of which is, by making woman believe she is indebted to us for protection against man's barbarity, to cause her so universally to side with us, that it is all but impossible for any of those infidels to connect himself with the beautiful half of humanity, without entering into partnership with one who will make him fork over the needful into our treasury, and insist, furthermore, that their children shall be educated after their mother's fashion.

Because we have prostituted organization and power
to the infliction of evil, our opponents think a leader
must necessarily be a tyrant, that organization must
unavoidably oppress individuals, and that the power
which leadership and organization can alone create, must
inevitably be abused. Thus, they are self-preventive
from doing us any harm.

Napoleon thrashed his foes into a knowledge of how
to conquer him, but common sense can't be cudgelled into
nothingists. [*A priest. Of course not. Did not we
maim their brains, as we shall those of their children,
in spite of all they can know as to how to prevent it ?*]

'Tis clear our enemies are ignorant of the A. B. C.
of the science of power—of the art of success. [*Bursts
of contemptuous laughter.*] Figs and fiddle-sticks for
the infidels, reasoners, retailers of disgregated facts, isolated truisms, and such like negative, barren, and resultless things.   Here is a specimen of their logic : —

"You must tear an old house down, and clear away
the rubbish, before you can build a new one on the same
site."

Who, with a spark of common sense, don't see that
the case they cite signally fails to be analogous to the
one in question ?   Would it be advisable or necessary
to tear down all the houses in the world, (for the case to
be covered is world-wide) before building any ; nay,
before even knowing how to build any ?   Would not
people rationally ask—" If you tear down our old houses
what better ones will you give us in their stead ?   And
why not give us a specimen of your architecture before
exposing us without *any* protection, to the scorching sun
and pelting storm ?" [*Laughter.*]

Well, every thing is now so entirely to our satisfaction that henceforth our easy task must be to keep
matters exactly as they are.

Democracy apparently don't expect government to do
much good.   To prevent it from doing harm is nearly
all that is attempted.   Lest the " necessary evils "
democrats sullenly submit to, should have time to

cement their power too much, very rapid "rotation in office" is resorted to. But the dear people have not made the time their chosen humbugs and privileged swindlers hold office, too short to enable each to grab his handful of feathers from the public goose. [*Laughter.*] Democrats ignore the great truth that it takes people a long time to do good, and but a short time to do evil; a long time to create, and but a short time to destroy. A temple it would take forty years to build, a few Vandals might demolish in a few hours, and millions of public funds, long and laboriously accumulated, may be Swartwouted in the shortest time to which democracy can possibly reduce the term of office.

Leaders could not possibly abuse any power, however great, which organization can bestow,—nor could they so organize the people as to make them, as they do now, destroy their own liberty and happiness — if mankind looked for results only here on earth. If the people settled with their sociological artificers as they do with lower mechanics. At the worst, rationalistic political moral and social architects could no more impose on those who confided in them, than do those whose artistic and mechanical skill is confined to the simpler and less complicated sphere of things more palpably material. If men would not attempt to "quit their sphere and rush into the skies," if they could gain but sense enough to see that the earth must be their final home, their leaders, however well organized, could no more become so powerful as, to any dangerous extent, to oppress the many, than an ounce in one end of the scale, could make the other end, with a pound in it, *and no string pulling at it from above,* kick the beam. * But we know how to keep the children of earth, constantly

* The words God and Spirit, convey no more meaning than so many letters thrown together at random. But admitting that, after abstracting matter and its conditions, there be an unconditioned residuum, transcendentalists themselves admit the impossibility of our knowing any more on the subject than the barren fact. *Can* it be of any more consequence to us than is the motion of those stars which

busied with matters purporting to be "beyond the skies,' [*Cheering.*] Thus only can we abuse, prostitute, and disgrace organization, leadership and power. Thus only can we obtain power of which the people who give it cannot divest us. Thus only can we convert the three requisites for the production of sociological science— leadership, organization and power, into bugbears, with which to frighten the well-wishers of humanity into do-nothings. Thus only can we perpetuate the idea that the very best government obtainable is the tom-foolery which *any one* elected from the crowd can practice, or soon get the hang of. Thus only can we keep men from finding out that *legitimate* government is the science which its professors can acquire only in proportion as they acquaint themselves with all the other sciences—that it is the science of teaching human beings how to live, on the average, four or five times as long as they now do, and in a condition happier than they can now accurately conceive of. Thus only can we prevent the many from gladly employing and richly pay- ing a few to devote their whole time to the performance of all this, and thus only can we prevent these powerful employers from compelling their sociological artizans to perform their tasks or give place to those who can.

But all this, by the crafty means just indicated, we do and must continue to prevent.

Oh, is'nt it rich to see solid, earth-made, sensuous beings, striving to promote a self-denying and therefore suicidal and monstrous species of liberty, virtue and honesty, conceived beyond the skies, but born and nur- tured on the earth? To see those who, but for us, might be as much happier than lower, animals as their reason- ing faculties are superior, making the superiority of their faculties the very means of degrading themselves below vegetables?

Oh, how luxurious it is to view the misery we create

have *no* parallax? Is there a shadow of probability that it is of as much, since the to us fixed stars visibly exist? Then why waste valu- able time, and exhaust brain strength on the subject?

in contrast with the happiness we prevent. [*Cheering long and loud.*]

And how do we keep the world gaping at what lies beyond human ken ?—and thus drag the intellectuality of the nineteenth century almost down to a level with that of the primitive ages ?

BY EDUCATION.

By the only means by which our opponents could, if they would organize under able leaders, with their whole time to devote to their business, dethrone us, and displace misery by happiness.

But our foes can no more overthrow us by attempting to teach reason to men and women, or rather those big enough to be such, than they can build a house by beginning at the top and working towards the bottom.

The meek friends and teachers of little children conciliate the women, defy the men, and subjugate the world. I tremble lest the pretenders to reason, and devotees of everlasting free discussion should learn common sense by our example. [*A priest, in a rather tremulous voice. Fear not, them. If we are ever subjugated it will be by foes of our own household. If any priest in high standing should ever succeed in turning the battery of education and organization against us from the fortress under his command, we might be done for. 'Til then, I repeat, fear nothing.*]

My friends, those who love freedom which they don't know how to get, and glory in that which may the Devil give them good of, delight themselves, every fourth of July, by reading over the Declaration of their fancied Independence. I will inform any Whited Sepulcherite, who may be here for the first time, that we, too, annually jollify ourselves over our Declaration of Independence—over the charter of our liberty to prevent the happiness and trample on the freedom of mankind. But before doing so, we will partake of our delectable

FEAST OF HORRORS.

I was in for it. Even a look of disapprobation would cost my life, I was morally sure. Just at this instant my career would have ended had I been observed, for I actually staggered with horror, on discovering, whilst the company were crowding to the supper room, that the clothes we had on, though as fine in texture and as beautiful in colours as any I had ever seen, and apparently differing in nothing from those worn in fashionable circles, were spun from human sinews and muscles, and sewed together with thread made of the heart-strings and lung tissues of beautiful young girls!

After a priest had, in due form, evidently from mere force of habit, invoked the blessing of Woollywolf, Chambaabee, and the Holy Boogeeboo, * —

The first course consisted of the brains of sailors, beat out with hand spikes.

The second course consisted of the flesh of slaves, roasted alive, or whipped to death.

The third course consisted of the flesh of laborers, worked and starved to death. Here the cook's art was perfectly marvellous. For despite probabilities, this course was more savoury than either of the preceding ones.

The relish with which these viands went down the throats of both ladies and gentlemen, I was obliged to imitate as well as my horrified feelings would let me. However, as I found them no wise different in taste from my every day food, I tried to forget, and managed to swallow.

But the worst was yet to come. For, after washing down this horrible meal in brimming goblets of human blood, tasting, however, like the choicest and costliest wines, in which the rich attempt daily to drown their enui and disappointment, —

The desert consisted of the flesh of the most sensitive of the human race, distilled through mental agony !

* The three omnipresent gods of Whited Sepulcherism.

Its taste, however, did not differ a whit from the most costly dishes served up at the tables of rich parvenus.

Oh, with how much less mental dissatisfaction could I have supped with a Cannibal Islander.

Whilst we re-entered the lecture room, a full band of music, played " Rogues All, " and " Ourselves, Right . or Wrong," which many of the company sung, and which was over and again encored.

A priest in full canonicals, now ascended the rostrum, and, opening one of the largest and most antiquated books I had ever seen, said —

My friends. The ponderous charter of *our* independence though every word, and letter, and point in it is worshipped by those it is the means of enslaving, is of such a nature that I shall be able to give you its whole pith in so few words, that instead of tiring of it, you will rather wish to hear it repeated, in order to laugh your fill over the joke we have played off on mankind, and, to cap the ludicrous, continue to play off, even on the nineteenth century.

The characteristic which distinguishes man from lower animals, is, his capacity to enquire whence he and all else originated, and how all will end.*

The knaves who naturally germinated from the primitive mass of fools thus answered this question :

Woollywolf, after spending the first half of eternity in hearing, seeing, feeling, tasting, smelling, and being nothing that we can know of, got tired of the monotony of the thing, or rather nothing, and made all which is.

---

* But for man's capability to ask a question more irrational than any lower animal is capable of conceiving, he never could have acquired even the fragments of science which will some day, by being joined together, formed into a whole, and made the base of sociological science, dethrone Whited Sepulcherism. By studying the mystical, man has discovered the phenominal. Seeking for natures cause, has introduced him to so many of her laws, that he will find out even the most complicated of them. The first knaves and primitive fools were mutually benefited by each other. Thus germinated leadership and organization.

In such a rough and tumble manner, however, that day and night occurred before the sun was created. The sun revolved round the earth, which was flat and stationary, and above which, the stars were set in a ceiling, the other side of which was a reservoir of water, which was sifted through this ceiling whenever rain was wanted.

The whole of this was intended, by its creator, for the eternal accommodation of a single pair of each species of animals, including man. Death was never intended to have entered into the world, even to accommodate lions and tigers; and never would have taken place, had our first mother continued blindly obedient. Woollywolf privileged her to do any thing she pleased, except to pull her husband's nose ; a thing it never would have entered into her head to do, had she not been forbidden; and for doing which, death was entailed on all living creatures, and eternal damnation on ninety nine hundreths of man kind. This damnation consisting in having the flesh of the soul eternally carded off with red hot hetchels.

From the incestuous intercourse between the children of the first pair, mankind, red, white and black, descended.

As might have been expected, those so abominably begotten became so corrupt, that Woollywolf, in a rage, tucked a male and female of each animal, and a single human family, into his breeches pockets, flew up above the reservoir with them, and let its whole contents down on the earth, destroying every thing which breathed, except the fishes.

After the water had dried up, Woollywolf descended to the earth, and emptied his pockets. But the means by which he undertook to reform his creatures met with such bad success that he would afterwards have drowned them all, had he not promised not to do so.

Failing to do any good by inflicting evil on his creatures, Woollywolf resolved to reverse his tactics, and try the experiment of inflicting it on himself.

Through the agency of the Holy Boogeeboo he

managed to get into the womb of the daughter of a rag-
picker, betrothed to a swill-boy, and was born in a hog-
stye, and called Chambaabee.

On which occasion, all the stars in the heavens, to
show they could be as small as their maker could be
degraded, came into his chosen residence, and paid their
respects to him and his mother; and even the Holy Boogeeboo consented that a messenger should be sent to
stop the boohooing of the swill-boy, and to assure poor
Jo that he would find his sweetheart a virgin, after all.

Chambaabee often used to weep over the miseries of
mankind. For about thirty years he led a vagabond
life, accompanied by other vagabonds, committing all
_sorts of petty crimes against decent society, destroying
hogs, stealing horses and colts, appropriating and tramp-
ling down corn, on the Sabbath day, girdling fruit trees,
upsetting the stands of market-women, and scattering
their apples, candies and lose change about, for one of
their number, who carried the bag, to pick up.

Sometimes he cursed the rich, up hill and down.
Then, again, he would commend the most exorbitant
usury, and laud the grossest dishonesty. He taught to
do good to all, and then to do no good thing, not even
to fulfill one's duty as a magistrate, except to avoid being
importuned.

He preached forgiveness, and then declared he himself would forgive no one who should doubt the precious
tale I am relating, or who should not believe in him and
his doctrines. Now, one half his doctrines were in flag-
rant and direct contradiction to the other half. What
was man to do? Of course, put himself, stone blind,
under the guidance of those who undertook to explain
them. [*Roars of laughter.*]

One doctrine, he so emphatically insisted on, that
there was pretty good reason to suppose he really meant
it. It was, that man should submit, without any re-
sistance, to whatever tyranny the most odious might
chose to inflict, and destroy all his earthly happiness,
even to emasculating himself, for the kingdom of Heaven's

sake. In his own day, he gained but few adherents ;
and finally, became so odious, that a mob of cow-boys
nailed him to a barn-door, when he squealed like a pig
and died. At that instant, the moon turned into a green
cheese and the sun swallowed it. This made the north star
roar out "crackee," so loud that the biggest barn in the
neighborhood split in two, and many dead horses were
startled to life, went into their old stables, ate up their
fodder, then harnessed themselves to their old coaches,
drew their old masters to church, and then vanished.

Chambaabee himself arose from the dead after a
while, went into a place where his disciples were assembled, said "peek-a-boo, old Grim can't come it over
this child," and then flew up to beyond the skies.

His followers eagerly looked to see if their old master
had left any thing behind him, and found a leather
medal, on which was a charcoal sketch of a stuck pig,
nailed to a barn door, with H. S. I., the initials of
Hypocrisy Superstition and Ignorance, over his head,
which the faithful now suppose means—Here Salvation
Is.

[*The priest here sat down; and after the laughter,
which was long and loud, had ceased, the president re-
commenced.*]—

My friends.— It would seem as though human
gullibility must have been tasked to its utmost by the
theology which underlies Whited Sepulcherism. But I
shall proceed to show that folly can go as much further
than this, as men can perform more than children.

Do but reflect that the precious stuff you have just
heard, is but the training the child's capacity to be imposed on receives, and then you will see, that the only
remaining difficulty must be to furnish nonsense sufficiently monstrous, and enough of it, to satisfy the man.
This, I have to confess even our inability to perform;
for the more we cram men with the very concentrated
essence of humbug, the wider do they open their insa-
tiable maws for larger gulps of what we palm off on
them for

LAW.

[Extravagant glee. Music. "I-A-W. Law."]

Of all the insulting humbugs—of all the inflictions for which mankind have to thank us, and, for which they most heartily *do* thank us, there is not one, with which we have so much cause to be satisfied, as with the law.

So long as we can keep our slaves hugging this chain, we are perfectly secure.

If monarchy deals out law with a spoon, democracy piles it on with a barn-shovel. Man seeks law and liberty in the farthest possible remove from both.

Oh, is'nt it rich, to see the good people, clinging for dear life to the huge mass of confusion we keep piling for them; and fearing if they let go their hold on this, they will sink into anarchy. They suppose the law is the rule for them to go by; but the instant they try to make the slightest practical use of their guide, they find they can no more understand it, than if it had been written in Chinese. Besides, the rules of their every day life are contained in books it would take them all their lives to read; [*Laughter.*] which books, even lawyers don't *pretend* to understand, 'til after long years of hard study.

Rise, Ladies and Gentlemen, and thank the priests of Woollywolf, here present, and through them, all their brethren. [*All rise, and bow to the priests, who receive the honour sitting.*] 'Tis they, who teach the dear people, the good people, the respectable people, the useful people, [*Laughter.*] to expect nothing but buffeting and misery " here below," in this unalterable "vale of tears"— to love those who hate, despise, *use* and humbug them as we do, by means of what is palmed off on them for law, and who unfit them, in youth, to know the difference between *real* law and the *sham* we give them for it.

This law may be compared to a bed of roses; so fair, so sweet, does it appear, spread out by our most worthy Blackstone, Kent, and others. The world has been lured to repose thereon. But oh! has it not felt the thorns?

So long as we can keep mankind under sham law, we can tax them to our hearts content, rob them as much as we please—in short, do with them whatever we like. What anarchy—what confusion, have we not inflicted on the folly-ridden world in the name of law!

That law and justice are two different things, is proverbial even among our victims. But this great truth, though pronounced by themselves, makes no practical impression on their theologically bewildered brains.

But in order to enjoy a full view of the evil we have inflicted on mankind by *sham* law, let us take a view of what might, and but for our effects would, be law. For we cannot fully enjoy the false, without seeing it in contrast with the true.

The material, is the base of the intellectual. (Not that there is any such *entity* as intellectuality, any more than there is such a *thing* as attraction, electricity, light, or darkness.) Therefore, *true* law in intellectuality, must be as spontaneous, as consistent, as harmonious, as *mathematical*, as it is in materiality the most palpable. We can no more *make* laws for intellectuality than for materiality. We can but discover and define them. By discovering what law is, we might avoid the folly and consequent misery of attempting to compel people to act, contrary to it. We here see that the very *ne plus ultra* of anarchy is produced by attempting to govern mankind by *made* laws. Now, almost every title of what we have put off for law have we not cause to be *made*, and that, too, by those who either had no conception of *true* law, or, if they had, were too *smart* to make known their thoughts on the subject? [*Here the company, (ladies excepted) kicked over the benches, rolled on the floor, and held their sides. The men shouted, and the ladies screamed in extatic delight.*]

External force is as inapplicable to moral, as to physical law. I have half a mind to tax the world the expense of strapping the earth with iron, to keep its centrifugal force from throwing it to pieces ; or to make the goodies pay us for pulling the limbs of their apple trees to make them grow faster. But no. That would not make mankind any more ridiculous, and not half as miserable. Besides, don't we now tax the dear people all they can possibly pay ?

The charlatanry which *makes* laws, can never be displaced by the science which could discover them and their connection, 'til mankind are, *from infancy*, educated in the phenomenal and real, instead of in the miraculous and imaginary, and until their leaders are materialists instead of supernaturalists. Our opponents have thus far, with but few exceptions, been little more than mere faultfinders, who grumble and fret, and boast of their reason, and talk about truth, and retail disconnected, isolated, and therefore all but useless facts, yet lack the very rudiments of common sense. They don't know enough to organize, to develop the faculties, both cerebral and muscular, of children, and to teach them to reason and understand, ere we, through organization, unfit them for doing either, and render them physically contemptible, and incapable of knowing truths, of any magnitude or intricacy, from falsehoods. Ere we indelibly fix in their minds the grossest and most pernicious falsehoods, as the most precious and important truths.

Oh, how delightful it is to see our foes distrust and abandon all who attempt scientifically to do any thing to build up the cause of humanity. All the genius, learning, and *talent*, are thus, however unwillingly, driven into our ranks. Even science has to bow to, shape itself by, or at least coquette with, our theology. [*Roars of laughter.*]

Only through a knowledge of connected science, can law be defined. But what of this vast and difficult study, which requires such elaboration, can those know, who, in order to obtain a living, have always been, and, always must be, more or less confined to some speciality? What of science do those who make laws know, except the science of getting elected to take a turn at swind-

ling, robbing, and humbugging their constituents?
[*Laughter.*]

To manage some pretty shop; to build a ship or
house; to make shoes, or even to cut hair, the good
people can see, requires time and teaching. Yet those
who are to determine what law is—who should discover
how to make man happy, free, and good, need only to
be born to the business, or *elected*, often by fraud, sometimes by violence, and always by humbug, in order to
be fully qualified! [*Prolonged laughter.*]

Whenever our victims get into difficulty which they
have recourse to law to settle, they turn to the ponderous books to see what law is. If they are not labouring
under some maniacy, they perceive only that they might
as well have consulted a brick. Those of their friends
who have experienced what law is, always advise them
to desist, let the case be what it will; If this advice is
not taken, the devoted victims deliver themselves blindly
into the hands of those who pretend to understand law.
After a year or two of the most torturing vexation and
anxiety, attended with ruinous expenses, would be grave
looking judges * pronounce a verdict which they pretend
is in accordance with the meaning of some particle of
the opaque mass they have been looking into for light
on the subject; and although this appears such an
astonishing mental feat that its performers are considered
prodigies of wisdom, the people imagine this same opaque
mass of law is *their* every day *guide;* and that without
it, they would be in a state of anarchy! [*Great glee.*]

How clear constitutional law, the most important of

---

* Democracy seats on *benches*, an army of despots, invested with
power more insultingly tyrannical than any throned autocrat dares exercise. Heads must be uncovered before them, and whichever attorney
can manage to put in the toadyism "your Honour" the thickest, is
most sure of winning his suit. Respectable ,and worthy people have
been imprisoned for contempt, often unwittingly committed, of some
miserable devil popular suffrage has fished out of the mud gutter, or
made a judge, just in time to prevent his becoming a hall-thief or pocket-
book dropper.

all, is, may be inferred from the fact, that no two judges, of contrary politics, ever construe it alike.

We give the law some appearance of being in accordance with right, else it would be no go. Thus, the booby who should see a piece of loadstone riveted to a piece of steel, would think their attraction for each other greatly assisted. Only the scientific could see that the real law of attraction could thus alone, be completely destroyed. We cannot rivet the sexes quite so inseparably as we can iron and loadstone, but we do our best, and proportionably destroy their mutual attraction for each other, whilst seeming to strenghten it ; and, to the utmost extent to which it can be done, promote the grossest lust and sensuality. [*Cheering.*]

The marriage law is professedly aimed against prostitution. Yet prostitution is acknowledged, by the marriage governed community, to be an unavoidable, "necessary evil;" of course, a part of their system. Marriage can no more repudiate what it has never lacked than a fish can deny his tail.

But how much of what is called marriage, is *really* any thing more or less than *wholesale* prostitution? We must keep marriage what it is ; if this fortress is ever yielded, our whole scheme will go to ruin.

Respectable society is afraid it will be impoverished by having to provide for too many children, unless parents are married. They don't see that those least capable of providing for their children, produce far the greatest number, *through marriage*, to fill alms-houses with, or go begging through the streets. [*Laughter.*]

Black children, whether their parents are married or not, are worth $300 as soon as born, although to be brought up so as to make them as useless as they possibly can be. Yet *society* don't know how to educate white children, so as to make them other than an expensive burthen. [*Prolonged glee. Music. Air "Popular Ignorance."*]

We have *made* laws ostensibly to promote the payment of debts. How could any thing appear more just?

Yet that these precious statutes are the cause why such a vast amount of debts are never paid, we see clearly enough, when we look at the fact that gambling debts, which the law requires the non-payment of, are with scarcely an exception, scrupulously paid.

Nearly all the debts collected by law, are court fees and lawyers charges. The reason of this is, men do not now, in reality, trust each other, they do but trust the law; and when it is most convenient, all things considered, not to pay, the creditor is at liberty to get all he can by, through, or out of, the law. [*Laughter.*] The debtor makes oath that he gives up all his property; and, after getting his liabilities out of existence at from 5 to 20 cents on the dollar, keeps a coach and servants, and lives in upper tendom. [*Cheering. Music. Air.* "*Ain't We Smart ?* "] . If any body dares insinuate that such an one has committed perjury, he had better look out for the entanglements of the libel law.

The poor creditor only gets laughed at or blamed if he complains. What right has he to find fault ?—To annoy and persecute a peaceful, law-abiding citizen ? What more could be expected of one than that he should fulfill the law ? [*Roars of Laughter.*]

If any one should *take* from this good law-keeping citizen, without paying for it, ever so small a portion of the vast amount *kept*, without being payed for, the law would consider the taker a thief. Yes, let twenty-five dollars be *taken*, unpaid for, from half a million *kept* unpaid for, and the bold *taker* is, by law, a criminal, deserving to hammer stone for long years, without pay, and to lodge in a stone cell three feet by six ; whilst the treacherous, cowardly, sneaking, perjured *keeper* of the half million from which it was *taken*, dwells in a palace, goes where he pleases, enjoys all the luxuries of life, and stands more than an even chance of being elected to some high and lucrative office, by those who laud nothing (our holy religion excepted), so much as *smartness*, and who honour wealth, however obtained, more than any thing except our Gods.

But the most exquisite concentration of infernalism, is the operation of law, when working people, and particularly sewing girls, ask pay of their employers, for the labor they of necessity have to trust for. Many of these employers, knowing that the law is *all they have to fear*, order working girls, out of their stores without paying them, as soon as they have deposited their work, although they know these poor girls have not the means of procuring a lodging for the night. Yet these same employers would not dare sit, whilst one of these same girls was standing, in a car or public saloon, for fear of being hissed by the public, who quietly allow innocent and industrious females to be starved or driven to prostitution, because *they can have recourse to the law* in the matter of pay. They interfere if a lubber attempts to make a female stand whilst he sits, only because there is no law in that case. I verily believe, that if the law required people to eat three meals a day, they would do all they could to avoid eating more than two ; and that if a law was made to enforce the civilties between ladies and gentlemen, the coarsest boorishness would soon reign in the politest circles,

Unfashionable mercenary prostitution not thriving quite so well as we wished in the state of New York, we procured a law for the preservation of chastity ; since when, we have had no cause to complain. In the space of three months, during one of our business suspensions, fifteen hundred chaste girls, in the single city of New-York, became food for the most loathsome disease our efforts have ever inflicted on the human race. Thousands of dollars, by means of this law, are annually extorted as compromise-money, from beguiled men, by females of uncertain character ; whilst wily debauches, who before dealt mainly with lose females, now ply their arts among those who would sooner die than appear in court against them.

More than three fourths of the human family who have arrived at the age of maturity, we manage to keep quartered on *real* producers in unproductive idleness, or worse than unproductive, because destructive, or duplicate labor. The bottom of the ocean, by means of *destructive* labor, is covered with the products of productive industry, and unnecessarily strewed with human corpses. [*Music. Air.* " *Free Trade and Sailor's Rights.*"]

The substitute for justice called charity, and to a great extent regulated by law, more than any thing else, fills .our streets with beggars, and our alms-houses with paupers. By means of charity, we kill three birds with one stone. We degrade a very large portion of mankind below all hope of redemption, pile another most oppressive burden on producers, and cause the most exquisitely villainous speculation. Oh, what enormous sums, raised for the perpetuation of pauperism, stick to the fingers of those intrusted with their disbursement. But we have, probably, no cause to rejoice at this ; for the money stolen, undoubtedly does less hurt than it would if honestly laid out for what it was intended, Scientifically managed, a tithe of what is now spent to perpetuate human degradation, would drive it from the world. [*A priest. " Ay, and such a beginning would probably be followed up by that which would eventually drive us and the Whited Sepulcherites from existence. But we'll see to that, fear not.*"]

I might go through the whole list of the laws we have caused to be *made*, and show that, without exception, they produce the most pernicious effects, and generally cause results exactly the reverse of those ostensibly intended. [*General laughter. A voice. " Mr. President, do you think the youngest of us will live as long as it would take to go through with a list of all the laws which have ever been made?*"]

Well, then, let us amuse ourselves with briefly detailing a few of the practical workings of these laws.

Not five per cent of what the law considers murder is ever detected ; about one in ten of all the hanged know nothing respecting the perpetration of the crime for which they suffer ; more than one half the victims of the gallows are *morally* innocent, and but an infinitesimal particle of what really *is* murder most foul, does the law consider such.

The same may be said of what the law considers theft, robbery, forgery, perjury, or any other crime. The law sanctions the perpetration of rape, if compelling females to embraces they loathe is rape ; and crimes against persons are hardly ever punished, though the unhappy sufferers by, and witnesses to, them, are frequently imprisoned along with murderers. Yet consider the vast sums the dear, good, respectable people pay for building jails and prisons, and for the support of armies of police officers and prison guards, and keepers. [*Hear. Hear,*] But the wholesale slaughterers, the almost exterminators of mankind, those who indirectly clip three fourths or more from the length of human life,* and spoil, with sickness, the remainder, the law to prevent murder takes no cognizance of. Military adventurers. Swill milk dealers. Distillers. Quack medicine venders. Food adulterators. Air-tight car, lecture-room, church, and theatre builders. Corset makers. Pastry and other cooks. Tobacconists. The manufacturers of almost every thing, *in the manner in which it is now done.* Morning news paper proprietors, many of whose employees linger, the little while they keep *some* life in them, under ground. Such are *not* murderers, but on the contrary, very respectable and worthy people, *according to law.* [*Prolonged laughter.*]

Imagine a coroner's jury, composed of swill-milk dealers, consulting a day or two over the body of *one* dead infant, to determine whether or not its mother ought to have her neck broken for having, rather than sacrifice the value of her whole after existence, strangled it as soon as it was born. This is vindicating the *majesty* of the law. [*Great glee.*)

To pass a piece of paper which is worth *nothing*, for whatever value may be written or printed on it, is con-

---

* It has been scientifically shown that man now gets but a tithe of his rightful and possible existence.

sidered forgery in law, and highly criminal. But we procure to be performed, in that line, and under the *special* sanction of law, a species of forgery to which what is *called* such is mere peccadillo.

The counterfeiter unsanctioned by law is not forced to cheat the world by his false tokens, unless he pleases. But we force all the world to *cheat* each other by means of tokens which are so nearly allied to valueless ones that the difference is but infinitesimal. If all the so called counterfeit bills in existence were presented, for payment *at once*, there would be *no* gold or silver—*no* real money, with which to redeem them, 'tis true. Well, if all the *good bills* [*Laughter*] in existence were presented for payment at once, together with all the *true* notes based on them :—what per centage of real money would they produce ? Barely enough to make the difference between the smallest particle and nothing perceptible.

But this is not half the villainy of respectable, law-sanctioned forgery. By inflating the currency to the verge of worthlessness, a system of credit is introduced which inflicts on producers an army of superfluous distributors of wealth, who convert trade into a species of gambling, fraught with consequences infinitely worse than so called gambling produces, and ending in universal bankruptcy every ten or fifteen years ; suddenly and *totally* depriving producers of their pitiful means of keeping sensation in their toil-worn bodies.

Goods are trusted out at a profit calculated so as to make all who *do* pay, the tools for sustaining in unproductiveness a very large army of those who, dishonestly will not, or *unluckily* cannot, pay.* So the would be honest trader, if he has *any* brains, except in the organs of *smartness*, reason thus with himself :—If I pay, I shall contribute to the most vicious system of swindling and oppression of labor which it is possible to invent, besides

* Well-kept statistics show that 85 out of every 100 traders fail. That is, 85 out of every 100 distributors of wealth are burdensome impositions on producers.

conniving at self-swindling. If I do not pay, I shall feel like a sneak, and I hardly know which horn of the dilemma to chose. To such a wretched alternative does the law reduce those who *would* be honest if they *could.* [*Prolonged cheering.*]

When the grand failure comes, however, no one is expected to pay unless it is perfectly convenient so to do. Every one holds on to what he has got trusted for. Even the few who do pay, had better do so on the sly, lest their unfashionable conduct should seem a reproach —an insult—to the majority of their brother traders. And if any one actually gives up all he has to his creditors—if a trader practices but the same honesty which the law *requires* of thieves—he is laughed at as an idiot, and those he so scrupulously pays, consider him so lacking in business capacity, that they are very shy of ever trusting him again.

A thief, is required to do without food, clothing or lodging, rather than *take* any thing without paying for it. But the man who can add perjury to the crimes of treachery, ingratitude, and breach of confidence, may *keep* untold wealth without paying for it, get elected to a judgeship, and be surrounded by beauty, fashion, and respectability too refined to keep company with those their idol has impoverished. [*Cheering loud and long.*]

Smart traders, say those who trust them, sell a great many goods for us ; on which we have made such roaring profits, that when they fail, if they don't pay more than ten cents on the dollar on what they *then* owe, we have still made roundly by their ability and *sharpness.* But as for the idiot who impoverishes himself to pay all he owes, he is too honest to lie, and consequently sells but few goods; and it is wonderful that some of the sharpers had not lighted him of every cent before it reached our coffers, So we'll not trust property in *his* hands again.

Permit me to refer again to our master stroke of policy.

We keep science in disconnected sections, and, above all, strictly confined to the most palpably material. There has been but one man, among the friends of humanity, who discovered that science was a unity, or rather that the different sciences were units of a great whole, extending from the simplest physics, to, and including, the most complicated—sociology. Which units, he discovered the connection of, so as to *form* this great *whole*, and by that means, drew thence a doctrine, which, if it were known to, and understood by, the leaders of mankind, and adopted, would banish Whited Sepulcherism from existence, and perfect humanity, physically, intellectually, and morally. But the great positivist— the organizer—the scientific, moral, social and political architect, is dead, and with Skrachfyre we hope. [*A portrait of Auguste Comte was now produced ; and, to test whether or not there were any traitors in their own ranks, all the company were invited to trample on it ; the priests eagerly scrutinizing whether every one did so with a will.*] Those only who have a well balanced brain, and some knowledge of the technical terms of science, can read his work understandingly. And if any one should attempt to do so except for his own private amusement, he would soon find himself spotted as the enemy of all good. [*Prolonged glee. Music. Air. "Folly Rampant."*]

Science, as now, in sections, and scrupulously excluded from government and morals, plays almost wholly into our hands. For instance :—The great science of chemistry has brought into use a thousand deadly drugs, which are almost universally supposed to be health restoratives. [*Cheering.*]

By means of chemistry, the largest portion of the material for the staff of life is converted into liquid damnation, which is sold for forty cents a gallon, to be shipped to France, whence it returns in the guise of Brandies and Wines which retail for from four to seven dollars per gallon. [*Immense applause. Music. Air. "Oh the Whiskey."*]

By means of chemistry, almost every article of food is adulterated. Even milk is drugged; or, better still, produced from cows diseased by distillery slops. [*After the cheering, which was uproarious, the music struck up the air "Swill Milk and Paregoric," and all who had tuneful voices sung an anthem to pestilence, commencing—*

"Dig more graves for the tiny dead."*

If mankind are ever educated in the phenominal, *connectedly and systematically*, their leaders will draw thence a doctrine exactly the reverse of that which has resulted from the study of the miraculous and supernatural. It will be discovered that the world, instead of being necessarily a " vale of tears " a " fleeting show for man's illusion given," contains all the raw material necessary for producing a paradise infinitely superior to any thing men, in their now gross condition, are capable of conceiving with any degree of clearness. With man and his environment perfected to the utmost capacity of *science and corresponding art*, undue want would cease, and all unnecessary misery (and nearly all misery is unnecessary) would end. Of course, vice would find no stimulous or inducement, and crime would be no more. [*A priest. " But we'll see that such an education shall never be given. We'll keep men's eyes so riveted on Heaven, that they'll never know enough of earth to dream of the possibility of materialistic morality or law. We'll stun their senses with our holy mysteries, as hunters stun the acute sense of hearing in partridges, with drums and dogs, so that the poor bewildered birds remain as if spell-bound on their perches, 'til they be-*

---

* Of all the children born in cities, more than half die before they are ten years old. We bow submissively to the will of our heavenly Father, who both " gives and takes away," and enquire no further into the cause. We don't even ask why our Heavenly Father thus afflicts only those of his creatures who belong to the race of his worshippers! Animals, below man, especially wild ones, have few diseases; and death, except by violence, among the young, is all but unknown.

*come the prey of those who could not have approached
them by other means.*"]

I tremble when I see men willing to pay for the
science, time and attention, it takes to make horses elegant and useful, and even conceding that to make these
animals as valuable as they are capable of being, requires
a knowledge of so many sciences, as often to absorb the
whole time of those who thus dignify horse-education.

How is it possible, that, not only the multitude, but
those who profess to be reasoners, can remain blind to
the glaring truth, that to perfect man's liberty, goodness,
and happiness, requires *his* perfection, and that of his
environment ? and that this can be achieved only through
science, which requires the time and attention, which can
be but very imperfectly devoted to it by those engaged
in ordinary pursuits ?

Negativists think they have achieved perfection, in
getting rid of the coarsest portion of the errors we fasten
on them in childhood. And, indeed, this is rather an
uncommon mental feat, and exhausts all the power we
have left the mental faculties, in most instances.

The man who makes the perfection of horses a scientific profession, is listened to with respect, and without
suspicion, and willingly and bountifully paid for devoting
his time to his calling. Whilst those who dare hint,
at making the perfection of man, *on earth*, a *scientific*
profession, are immediately suspected of being "aristo-
crats," unfriendly to self-government. It is mean, insulting, and venal, for *them* to think to make a living by
their profession. "Aint the people capable of governing
themselves in this free country ?" Yet men would consider it the height of folly to have the educators or gover-
nors of horses elected from the crowd, by a majority of
votes. They can readily see that such a method would
degenerate the best bloods to scrubby, worthless mus-
tangs, and keep them such.

Why do men not see that the election method keeps
them the miserables they are ? Why can they not see
that science is as requisite to *their* perfection, as to that

of horses, cattle, and even vegetables? and that if they cannot, whilst engaged in other avocations, attend to the science necessary to perfect horses, how much less can they thus attend to the science necessary to perfect themselves?

In short, why must horses, cattle, and even vegetables, be honoured with scientific leadership, whilst man is degraded to being led by the nose by low, swindling, tricky, ignorant demagogues? [*A Priest.* " *We know why.*"]

But vegetables and beasts have only to be perfected materially and physiologically, suggest some bipeds, to whom I daily fear some one will *effectually* reply:— make man and his environment as perfect as possible by means of science, and then, how *can* he be other than morally good and virtuous? And if man is not, like every thing else, to become as perfect as it can be theoretically shown he is capable of becoming, does he not belong to the lowest, instead of the highest, order of existence? [*A priest. I should'nt fear to have all you have said proclaimed on the house-top. My brethren are needlessly afraid of infidelity. All which reason can suggest to theologically-trained, forty-year-old babies, will be as resultless, as attempts to straighten a forest of crooked trees.*"]

We have converted an immense number of the human race to chattel-slaves, and made all the hired laborers so miserable, that it is impossible to decide which have the most uncomfortable time of it,—the wages-slaves, or the chattel-slaves. We contrive, also, to provide both species of slaves with a good share of mental misery, in the shape of envy. The poor never dream that the rich are as miserable as they are wretched. That if the poor are a prey to want, the rich are tormented with care, anxiety, and killing en-nui. If the miserable rich, refuse, at any time, to pay all we demand for keeping starving desperation from assailing them, we can let them be robbed, and share the plunder.

The poor are ever watching an opportunity to mob and despoil the rich ; nor, without our aid, can they always be conciliated by soup or alms.

In concluding, let us concentrate our delight, by reviewing our success-crowned efforts at a glance :—

We have made those science *would* redeem, do all in their power to kill their saviour in his infancy. They mock and buffet him, and whenever he attempts to show his *head*, with what sharp thorns do they crown it ; and put to open shame, him, who would gather and combine them. We have made man not only as miserable as he can be, but so all but totally depraved, that he not only puts his saviour *hors de condition* to do him any good, but rejects him for the veriest Barabas, the merest sham, the *ne plus ultra* of an impostor. Such an impostor, that he promises man happiness only after death. Such a sham, that the best thing he professes to be able to give them in this world is Democracy. [*Laughter loud and long.*] Yes ; free *elective* government. And how comfortable a thing this precious freedom, which cost so much blood to get, and requires so much treasure to maintain, is, we may see from the fact, that if Monarchy gives man a sprinkling of *law*, Democracy deluges him with it. If, in Monarchy, law is inflicted by those who arrogantly claim the right to a monopoly of its *manufacture*, in Democracy it is manufactured by those who receive their patent from the lowest, most ignorant, and most vicious of the populace. In short :—

We have extended from the remotest antiquity into the middle of the nineteenth century, a low, barbarous, and degrading superstition, in monstrous contrast with, and antagonistic to, material science, and preventive of, sociological science. We thus keep man as contemptible, ridiculous, miserable, and depraved, as he possibly can be, though surrounded by, and even a part of, that nature, which, if looked to, and depended on, and scientifically developed, instead of being despised for phantoms " beyond the skies," would make him perfectly free, perfectly good, and perfectly happy. What can concentrated evil—Whited Sepulcherism, more ? *

* The abuses of civilization and science, made even Rousseau sigh for primitive simplicity, and its fancied happiness.

The road from savageism *through* civilization *to* positivism—lies through mystery and slavery. Man, 'til he reaches that perfect state when right will be its own incentive, needs, and will have, the spur of bodily misery, or the equally sharp and impelling goad of mental anguish, and ennui, to force him on to his destiny.

Cobbet, in his History of the Reformation, has shown that the mass of the people are more miserable under Monarchy than they were during the middle ages, when governed by theocracy or sacerdotal despotism. Other things being equal, mankind are as much worse off under Democracy than under Monarchy, as the tyranny of a few is more onerous than that of the majority. The fact. that misery and ennui increases, as civilization advances, is so apparent, that superficial thinkers abandon their hopes of man on the very grounds on which they ought to found them. Misery and ennui spur man on to where he will be free from them.

But if our half civilization produces miseries unknown to savageism, this does not prevent its also producing pleasures unenjoyed by man in a state of natural *simplicity.*

Do philosophers know what they mean, when they laud, and sigh after, that nature from which man has *departed ?* I admire, prospectively, of course, that nature, to which man is *to arrive,* guided by nature's accomplisher, science, itself but nature.

I have seen, and studied, man, in his primitive, wild, simplicity, and found him, in the main, but as savage and cruel as he was ignorant; and were this not so, ignorance would be on a par with knowledge. True, civilization has, as yet, not *much* improved man, in the main ; but its work is not *accomplished ;* consequently, the result is only perceptible to the eye or the scientific seer.

Savageism is a condition scarcely worth living in; and if it were ever so desirable, man can no more return to it, than the youth can go back to manhood. Each time Democracy becomes so unbearable that man retreats back to Monarchy for respite, he touches the latter lighter and lighter. There are some oscilatory exceptions, but this is the rule. By and by, revolution will follow revolution in such quick succession, that man will gain little repose in Monarchy, which he will finally not touch, and have *no* alternative to the despotism of the ignorant and licentious majority but *positivism.* which alone can make him free ; and up to which. he will be, with few exceptions, and some short intervals of respite, but more and more wretched, and more and more a slave.

Democracy, cut adrift from the merit which circumstances often force to found it, and fully developed, is the tyranny of the necessarily ignorant many. Theocracy is the germ of leadership. Monarchy faintly shadows forth that leadership should be scientific. But Democracy puts the tail before the head, has no type in nature, and is a lie

I had all along observed,˙ that the satisfaction the
orator drew from his contrasts, was not unalloyed. Also,
that his auditors showed signs of not feeling quite so con-
tented and easy as they wished to.

Instead of demonstrations of delight, a look of despair suddenly darkened the countenances of both speaker
and hearers, as when a tempest-freighted cloud hurriedly
obscures a summer sky. The president paused a moment, and then, in a voice changed from a very full to a
dismally hollow tone, thus resumed :—

But what have I confessed? Nature, in and around
man, contains all the means requisite for his perfection.
She will surely use those means. She possesses no
vain, no idle power. None she *never will* use. As surely
as exist all the conditions, though undeveloped,—all the
causes, though in embryo, necessary to the production of
any effect. such effect has been, or will be produced.

In Democracy, too, I plainly see, we have given man
a goad so unendurable, that it will spur him on to
quicker and quicker change, or revolution, and consequent
enlargement of his sphere, 'til it will bring him within
that of positivism. Yes, at each revolution he quickens
his passage through his perihelion, absolutism, and nears
more and more its outward verge, 'til finally he will entirely
clear both absolutism and Democracy.

We have done all but the impossible. We have
insulted nature even to the extent of attempting to de-

throughout. Its ministers are charlatans, quacks, and the most execra-
ble villains, robbers, swindlers and cheats. All the parties which com-
pose Democracy unqualifiedly affirm this of each other. I most heartily
approve their verdict, and on this most vital point, sincerely honour
their judgment.

The United States, owing to its unbounded natural advantages. can
stand the tyranny of Democracy a great while. England has established
such a counterbalancing status quo between Democracy and Despotism,
that quiet is almost as stereotyped there as in India by the castes. But
the quiet of England is on a much higher bed, and the turbulence of
popular rule is, though almost imperceptibly, stealing over it. When
this gets the ascendant. Democracy will make the whole continent of
Europe caper to its lashing 'til it drives it into scientific government.
Then, farewell forever to mystery and demagogism.   .

throne her. Nay, we have *apparently* succeeded in sub-
jecting her to the government of a *supernatural* monster.

But this *must* end. The temple of natural, materialistic science already exists. Its walls are formed of the
simpler and more palpably material sciences, and its roof,
composed of materialistic moral, political and intellectual
science—Sociology, crowns it, and awaits but man's en-
trance.

All the capacities of nature *must* be developed.
What is in her womb must be born in spite of even us.

'Tis but a question of time when mystery must give
place to positivism. The order of the evolution of the
material sciences, has been discovered and demonstrated;
the theory of their connection is plainly and indisputa-
bly before the world; positive, moral and intellectual
science, it is now clear, is but their extension and perfection. The earth, instead of "*beyond the skies,*" will
soon be the object of man's affections, and the foundation
of all his hopes. Whited Sepulcherism is doomed!

The birth-qualification to govern, is worn out, and
long since condemned. Its Democratic substitute has
been tried over and over, again and again, and as often
found a complete failure. Political demagogues and moral quacks have alone profited by it, and it is fast getting
into worse odor than its predecessor.

Improvement, though it oscillates, has an onward
course, in the main; and, notwithstanding appearances
to the short sighted, there is progress.

Science is about to become a connected, organized
guide—a ruler. Nature, long in labor with the body of
man's saviour, gives unmistakeable proofs that his *head*
has undergone parturition. Wise men are preparing to
tender their allegiance to a king to whose government
there shall be no end; to present to a redeemed world, a
saviour worthy their eternal adoration.

Labor-saving machinery will not much longer in-
crease the splendid misery of the rich, and the squalid
wretchedness of the poor. Labor, capital and skill, so
long deadly foes, are about to enter into copartnership.
[*A fearful tiger for Fourier.*] The old, effete, morbific,

must pass away, despite our efforts to retain their festering rottenness, and all will become new. We, too, shall be abhorred, detested, hated, loathed, despised—oh, how much more intensely than we are now respected and loved. We, the scourges of our race, with none so "poor in spirit" as to do us reverence, must give way to its benefactors. Our names, associated as they will be with our deeds of darkness, will stink in the nostrils of the latest posterity, infinitely worse, than will those of our predecessors. For mankind's first deceivers necessarily taught their victims the rudiments of civilization. But we have done all in our power to prevent man from progressing beyond the mere initiatory of enlightenment.

Man's *earthly* redemption is inevitable. It might have happened in our day, had he, whose portrait we have just trampled on, been twenty years younger, and had a few capitalists been Girards. It will, it must come, whenever the leaders of mankind, or a majority of them, shall study and understand the crowning science which he has expounded, and without which, fractional science is nearly useless, and much of it pernicious.

Yes, materialistic moral, political and intellectual science—sociology—must sweep moral, political, and social charlatanry from the world. The towering pyramid of human folly and misery which we have reared must tumble, crushing and burying us in its ruins.

We had nothing to fear from science, so long as it existed only in parts, like the fabled temple of Solomon, ere it was put together. But the great architect came, who adjusted, fitted, and connected those parts, and of them built the walls of the great temple of science. Even this, we might have snapped our fingers at, for what useful purpose could a roofless building serve? Within its walls, mankind would not have been sheltered from our malice. We could there have deluged them with evil as easily as in the open field. Nay, those very walls have often furnished us with missiles for the devoted heads of those trusting to their shelter.

But the great master builder knew more than how

to put together the parts which others had prepared. He completed the temple of

SCIENCE AND LIBERTY,

[*Terific Groaning.*]

by crowning it with a sociological *roof*, beneath which, humanity will inevitably find a sure refuge from all our machinations.

Oh, let me not live to see the day when science shall include intellectuality—when moral shall be based on physical law, and when the expounders of the whole, shall be the leaders of redeemed mankind. Ere this, in earth's remotest centre, be my memory buried.

I have visited mad-houses, and heard the fearful ravings of the most hopeless maniacs. But never before had I witnessed any thing approximating to the horror, despair, and *mental* anguish which was depicted in the countenances of these demoniacs, during the concluding part of the speech of their infernal orator.

I have penned all I could make out of his concluding words. But he continued his attempts, even after his voice was so dry, husky, hollow and indistinct that it seemed to come through a throat and lungs, and from a tongue parched by the very flames of Hell.

At length he sank exhausted into his seat; his eyes showed only their bloodshotten whites; he breathed most laboriously. One instant I could hear his clenched teeth grit; then his jaws were horribly distended. His hands were now clenched, now spread as though the fingers were drawn backwards, and every muscle and chord visible, gave signs of the intensest agony.

The audience glared at each other like damned spirits in the infernal regions. Every moment I expected to see them clawing out each other's bowels, tearing out each other's eyes, gnashing each other with their teeth, or even, in the fury of their mad despair, venting their rage on themselves.

I could endure no more; so I tottered away, apparently without being observed.

PRACTICAL MEANS FOR THE EXTERMINATION OF WHITED SEPULCHERISM.

By exposing the wiles and abominations of a craft which, by means of a *savage* superstition, crushes the nineteenth century almost back to middle-age barbarism, makes moral cannibalism dainty, and every crime respectable by virtue of its magnitude, I have done mankind a service, for which I shall be sufficiently rewarded by my reflections, notwithstanding the consciousness of peril, should any ōf the innumerable spies and toadies of the infernal brotherhood find out to whom they are indebted for this turn.

True, the attention, during the twenty-eight years I have devoted to this task, might, otherwise directed, in all probability, have procured me the enjoyment of all the splendid miseries, fashionable vexations, disappointments, and ennui of an upper tendom palace. But still, I am self-satisfied, though in such a modest cottage that I am deprived of the inestimable society of parvenus, whose gaud and glare astonish man, *provisionally* captivates woman, and excites the envy of the million. But to my proposition:—

The mass of mankind *really* want tangible results. First, the useful. Second, the amusing. For these, they have always had a natural, common-sense, instinctive feeling of the *necessity* of depending on the few—on leaders possessed of knowledge beyond popular reach. Nor has even so monstrous and unnatural a thing as Democracy been able to eradicate this feeling, but only to suspend it, and for short intervals.

Science has not, until recently, seen *how* to satisfy humanity's great need ; and, naturally modest, it has not pretended to be able to do more than it was conscious it could perform. In the mean time, blind mystery has misled, and impudent quackery abused, the confidence which man has been impelled by a law of his nature, to repose somewhere. Mankind have been so long governed by unskilful or unworthy leaders, that even after they find they have been but wandering in error, it will take some time to get over their suspicion, that leadership *originated* error ; and to be convinced, that until scientific leadership was possible, error was their only alternative ; and being led about in it served the invaluable purpose of preserving organization and leadership, without which, instead of progressing even after they shall have found the right way, they would mope back to dark savageism. Thus, error and knavery have furnished the means—leadership and organization—of their own overthrow, and of man's scientific redemption.

Whilst the leaders of mankind remain the blind apostles of the unconditioned, the confidence the masses place in them must necessarily be blind and unconditioned.

But men will follow their *positivistic* leaders *understandingly*, and with eyes wide open *as to results.* The *what*, people *will* judge of. The *how*, they will leave altogether to their leaders, as they always have done in more palpably material affairs. The sociological artificer will differ from others, only in the encyclopedic magnitude of his calling.

Whilst men can see that single sciences require, for their elaboration, often a life-time, they think sociology, because it concerns all, *must* be of such a nature that *every* one is competent for it. They cannot suppose the " pedant they have placed on the throne of the universe " would allow that *something-or-other* which concerns *all* to be understood only by a few, and after " long and difficult elaboration." But many of the simpler sciences *glaringly* affect all, and yet are acknowledged to be understood but by those who make them their study, and often their business.

When men shall see that sociology—government, includes all science, and that science is but a headless trunk without sociology, they will realize how exquisitely they have been humbugged by the " *elective franchise.*"

The sociological artificer will be *required* to perfect,

as fast as possible, both human and external, physiological and physical nature.   In proportion as the material
and physiological is explored—understood, the materialistic—social—moral—*must* improve.   If all nature, including human, was but as perfect as it is capable of being
made by science, no vice, or sin, and but very little, if
any, evil could exist.   Death itself would be but the
natural decay necessary to *distinguish* or *manifest*
life.

All the while the *identity* of the sensuous being exists, such being constantly rejects, because it has fulfilled
its function, what it but just before cherished as its life;
and identity itself, will finally, when man shall live his
scientifically natural term, become worn out—tiresome,
and will willingly be parted with.

Those who have thus far read these pages, will no
longer wonder at the amount and enormity of crime, and
will see that only in proportion as Whited Sepulcherism
can be displaced by positivism, can "man's inhumanity
to man" be prevented, or but infinitesimally punished.
Except in proportion as organized science can displace
organized murder, rape, robbery and canibalism, can mankind gain any thing.

Yes, even Democracy—organization to destroy organization—life leagued with death to destroy life—leadership
to abolish leadership — and nonsense *systematically
taught*, is better than the stone dead *lack* of organization,
leadership, *life*, which negativism would produce.

To positivistic organizers alone must be intrusted the
task of ridding the world of the impostors and ignoramuses
who now govern it by means of flattering it that it gov-
erns itself through the elective franchise.   The praetical
question is, how are positivistic organizers to be consti-
tuted? and how are they to begin their work?   This,
the encyclopedic nature of positivistic sociology precludes
my doing more than hint at.   All that is now *certain*
is, that we have found a new and infinitely superior
world for the human race, the *minute* geography of

which exploration by its scientific discoverers and their successors can alone determine *as they proceed.*

Science, supported by a few enlightened capitalists, *might* begin the work of human redemption, and I will *opine* how, after one more remark :—Hard-shell protestants—infidels—serve a useful purpose, after all, which it would be doing them gross injustice not to acknowledge. They cause positivistic organizers to redouble their efforts, lest mankind should be disgregated back to savageism.

SCIENTIFIC REDEMPTION.

ACT I.

SCENE I. *A Sunday * Soiré in Upper Tendom, at which many intelligent poor, but no* MERELY *rich, are present.*

MRS. A.   How did you like the sermon, we heard to-day, Miss B. ?

MISS B.   Sermon, Mrs. A.? it was more like an oration ; or a lecture on science.   I could hardly believe I was in church.

---

* *The late clerical assemblage, in New-York, to promote the super-stitious observance of* " *the Sabbath,*" betrayed a consciousness of being engaged in not only a bad, but rapidly declining cause,   There was evident wincing and writhing under something " of great pith and moment " to be both said and done, without daring either.

Sir Matthew Hales and their backers, alack and alas, are all but phantoms, fast receding from the " aching sight " of these veritable Cotton ·Mathers.

One fact was incautiously stated at this meeting, which I beg Sabbathites not to let go in at one ear, and pass out through the other, without stopping :—

Sunday, in Paris and Vienna, is a day when labor *does* " breathe more freely " than in shops or churches: and as man needs it " *physically,*" he *uses* it thus.   Yet, in neither of those cities is crime so rampant, on Sunday, as in New-York, with its busines suspended, nearly all its places· of amusement closed, its brothels unlicensed and under legal ban. half, and sometimes all, its rum-holes shut up, but with ITS GOSPEL SHOPS ALL IN FULL BLAST !

Mr. A. And the new edifice, Miss B. ; don't you think the superiority of its architecture corresponds to the excellence of the preaching?

Miss B. Well, Mr. A., the hall in which the congregation assembled is perfect, and the whole edifice, to my thinking, is magnificent ; and oh, how completely and ingeniously ventilated. But what is the use of all those smaller halls, your munificence, and that of a few others, caused to be included in a church?

Mr. A. All will be explained by my colleagues and myself, according to promise ; or, failing to do so satisfactorily, we shall buy out the rest of the stockholders. Next Sunday, we are to have the use of the *whole* building, in which to make this explanation, *practically*.

Mr. B. Not long since, in this country, we might have been subjected to the discipline of the stocks, for holding such an unsanctified meeting as this on Sunday —beg pardon, *Sabbath* evening. [*Laughter. The company draw round attentively.*]

Mr. A. Well, every generation grows wiser than the preceding one. ˙Our pastor has evidently received a scientific, rather than a theological education. Ladies and Gentlemen, who of you would like to have Sunday devoted altogether to science, amusement, and rational education, both physical and moral. To have the budding intellectual faculties of children exercised and expanded by something, rather than contracted over, and to, nothing.

Nearly All.—I.

Mr. A. I thought so. I am an old church-member, and have selected and invited here, this evening, those of our communion whom I believed would second my views. We have had downward-levelling, destructive, "fought-bled-and-died" revolution, to surfeiting ; and with *such* results, that we need not fear to risk the experiment of upward, constructive, peaceful revolution. Besides, I am for adding another religion to the numerous onés already in existence, whose creed shall be founded on the known or comprehensible—the phenomenal—the positive. Whose priests shall make it their study to improve our *bodies* and their environment, and to amuse us, and instruct our children in the practically useful, and attend to their exercise, development, training and amusements. I have consulted a *certain* clergyman on the subject, and *know* that if he was assured of being sustained and furnished with the means, he would put himself at the head of the movement. Professors, capable of introducing science as a connected *whole*, and of informing us, from an *intelligible* point of view, *what* to do to be *understandingly* saved, await but a *call*. I will take $100.000 worth of stock to begin the enterprise.

Mr. C. And I, $100.000.
Mr. D. And I, $50,000.
Mr. E. And I, $75,000.
Mr. F. And I, $25,000.
Mr. G. And I,
Mr. A. Hold, gentlemen, 'til I can put down your names, and—but no, we'll have a special meeting for that purpose. I have no doubt, from appearances, our church stock will be a good paying investment. We will now see if we can make Sunday what it should be, "for man;" and *begin* to make the human race something more than the laughing-stock for all the rest of nature.

Mr. Negative. All this may be very well, as to yourself, Mr. A., and perhaps a few others. But in order to make the thing succeed, immense numbers will be necessary; and it will take a long time to enlighten their minds on the subject, to discuss the merits, and thus convince them of the truth of the new system, and make them perfectly understand what they are going about.

Mr. A. Understand?—convince?—enlighten?—truth?—Pray, Mr. Negative, *when* have the *masses understood*, or been *enlightened* as to the *truth* of whatever *system?* they have followed? Not the *working*, but the *work*, of sociology, can the *masses* understand.

Are they *understandingly* convinced of the *truth* of theology ? Are they *enlightened* as to the *truth* of Democracy ?
Or do they take both, but more particularly the former,
on *trust ;* supposing, hitherto only to be deceived, that
must be true which is backed up by such respectable
authority ? The *people* understand moral, political and
social science, the *summum bonum* of *all* science, Mr.
Negative ? why, they seldom pretend to be good at more
than *one* science.

But the more advanced society becomes, the more
important leadership *will* be, because the more scientific
it must be.

To dispense with leadership,—to make the sense of
the majority the test of wisdom, the *use* of science, in
sociology, must be denied ; or else, every individual, or
at least the majority, must explore the whole of it, as
it is a unity, and keep pace with its rapidly increasing
complexity.

But as science, and even leadership, are acknowledged
necessities in matters grossly material, is it not madness
to deny their increasing usefulness in Sociology ?—to
exclude them from the only sphere in which it is possible for the transcendent value of their functions to be
manifested ?

Let science have a chance to *finish* its work ; for, if
it fails, we can, at any time, have recourse to the bless-
ings of superstition, demagogism and infidelity.

Intelligent wealthy people support the church, because
nothingism is the only alternative. The sexual relations,
law, government, education, every thing ; even fractional,
crucified, headless science, is connected with, and more
or less dependent on, the pitiful caricature of organization which theology, in the last stage of decay, affords.
The ladies attend church as model artists, on which for
gallant gentlemen to feast their eyes, and delight their
imaginations. But the thing is beyond discussion's
reach, Mr Negative. We shall begin with a power of
respectable and admired talent, and a leader whom the
church venerates. Free discussion has become *almost*

as great a clog to progress as its suppression ever was. Popular liberty can amount to nothing more than rope enough for the people to hang themselves with, except in proportion as it is preceded by science, so extended as to include sociology ; and science, to be thus extended, must become *religion*; and be associated with splendour, melody, gayety, and all which the entire satisfaction of man's nature requires.

As fast as absolutism relinquishes, demagogism and social quackery occupy. Thus, popular sovereignty never can be a reality, 'til science is prime minister, invested with power over all, except man's right to decide on its merits *from its performances.*

This is the way *we* shall exterminate old fogyism, strangle both superstition and negativism, and dry up vain discussion. [*Folding doors open. Enter the learned clergyman, attended by a number of scientific professors, elegantly dressed, followed by forty or fifty ladies, mostly in ball costume, and a whole Sunday-school of little Misses and Masters. Music.*]

Gentlemen, please take partners for the Serious Family Polka. [*The clergyman selects Mrs. A. The professors select ladies who have been apprized of what was going on ; the children are eager and delighted ; some of the ladies and gentlemen are surprised, but passively yield to the influence of the current, and even Negative asks to be introduced to a partner, and is evidently under such strong convictions, that there seems every probability of his joining the church, without further discussion.*]

After the Polka, the first priest of *true* religion remarked:—

Ladies and Gentlemen.— Science and philosophy, never can achieve any thing like a triumph, 'til properly adjusted to, and associated with, the melodious, the gay and beautiful, and the amusing. Those who represent the gay, the beautiful, in a word, all that is enchanting, have hitherto been reduced to the necessity of lavishing their admiration, and wasting too much of their sweetness on the meanest of mankind, because such were,

more than any other class, associated with the music,
tinsel, and gayety, which the worthier portion of humanity have so unwisely ommitted, or so stupidly despised.

Does tender-hearted woman love murderers? Does
she admire those whose trade it is to make widows and
orphans? Libellous nonsense. She does, indeed, love
valour, but 'tis a virtue so little known to the soldiery,
that the *reality* of it would, most likely, stamp its pos-
sessor a *coward*. 'Tis seldom a soldier's idea of courage,
rises above furious recklessness.

Ladies admire soldiers, because they are gaily dressed,
jovial fellows, and associated with music ; and here is a
lesson of wisdom, of more *practical* use, than any we
have yet learned. Let philosophy and science not despise
it, 'til they are prepared to turn up their dignified noses
at the melody of birds and their gay plumage, the beauty
of flowers, the brilliancy of the rainbow's tints, and lovely
woman's all powerful influence.

What would a church, an army, nay, the world itself, be, without the melodious, the gay, the beautiful?
Most miserable and gloomy failures ; ridiculously incom-
plete and inefficient, like science and philosophy, hitherto.

Beauty, gayety, and melody, have too long been the
slaves of superstition, rapine, and murder. Henceforth,
let them adorn, enliven, and *complete*, science and phy-
losophy.

Next Sunday, the new church could not contain half
the people principle, expectation, and curiosity, had
brought thither. The pastor, the scientific college, and
the initiated, appeared in a costume of such elegance and
good taste, that, in it, the fine art of dress seemed to
have attained perfection. Also, they had made the most
of the very favourable circumstances under which they
were about to inaugurate a movement, the importance
of which, threw all previous ones into insignificance.
Elegance, taste, beauty, gayety, fairy forms in associa-
tion with music's soft strains,—all the requisites of

mirth, joy, and gladness, were there.   Victory was evident. For the ladies being completely charmed, enraptured, captivated; the men, of course, cheerfully and gracefully surrendered at discretion, and superstition was doomed.

The discourse consisted of a summary of what science, the religion of the known and comprehensible,. proposed to effect ; and the *use* of all the halls in the church was explained to such purpose, that although a few members sullenly withdrew, their places were more than doubly made good by recruits from other churches; and all Christendom, evidently, would soon have to join the great *reformation*, or find their superstition shops in a condition similar to that of the Mississippi flat boats, after steam navigation commenced.

Mr Negative remarked, that the religion of the *known* would be more expensive than that of the *unknown*, were it not that the *true* church will accommodate, in one grand edifice, the congregations which the *false* church assembles in six or eight superstition shops.

In the evening, the *first* edifice ever dedicated to *true* religion, was briliantly illuminated. In the large hall, "music arose with its voluptuous swell," and youth and beauty, childhood and old age, " tripped the light fantastic toe," and sung the requiem of superstition, demagogism and infidelity.   And this was but the prelude to man's destiny.   A destiny faintly conceived by the ancient seers.   A destiny which will realize the millenium and even Heaven of the multitude.   A destiny, the glory and happiness of which, however misconceived as to minutia, have not been over estimated, or too highly coloured, by the great sociological prophet, Charles Fourier.

The government of science, *as a religion*, must succeed wherever absolutism vacates, or demagogism and moral and political quackery will occupy.   Superstition has its root in the soil of savage ignorance, *wild* democracy.   Hence, in even modern, and somewhat *tamed*, democracy, superstition flourishes better than under *any* other form of government. In the United States and Mexico, it shows itself much more at home than in France, England, or even Rome itself. The worst tyrant never hated liberty. He only loved it so well, that, in his blind fury to secure it for himself, he trampled on the rights of others.

Those who think " the world is governed too much," have evidently not looked beyond the sham which passes for government. Of that, the least possible quantity is certainly the best. But of scientific government, we can never have a surfeit; for were we as perfect as possible ; were all phenomena understood ; leadership alone could *preserve* the knowledge which gives existence all its value.

THE END.

ROUSSEAU'S CONFESSIONS COMPLETE.

THE CONFESSIONS OF JEAN JACQUES ROUS-
SEAU. NEWLY TRANSLATED, WITHOUT OMISSIONS OR
EXPURGATIONS.

Period First relates to Rousseau's youthful adventures
to the thirtieth year of his age.

Period Second embraces his literary and public career.

Both Periods make two large, elegant 12mo Volumes,
sold separately, at $1 25 each, or $2 50 the set. Mailed free.

" There hardly exists such another example of the miracles which com-
position can perform."—*Lord Brougham.*

"There have been what purported to be translations of the world
famous Confessions of Rousseau before; but Mr. Calvin Blanchard's, just
issued, is the first that we know of which is unmutilated and accurate."—
Putnam's Monthly.

" It has been translated into every language of Europe; the librarian
of Napoleon devoted a large volume to the classification of the different
editions of it.— *Evening Post.*

" Blessed be the early days when I sat at the feet of Rousseau, *prophet
sad and stately as any of Jewry.* Every onward movement of the age,
every downward step into the dephts of my own soul, recalls thy oracles,
O Jean Jacques!"—*Margaret Fuller.*

The Confessions incidentally portray the remarkable
times immediately preceding the French Revolution. The
squalid wretchedness of the peasantry; the gross licen-
tiousness of the clergy; the gallantries of the nobility.
It introduces us to those famous philosophers, Voltaire,
d'Holbach, Diderot, d'Alembert, Hume; to Mesdames de
Warens, d'Epinay, and the enchanting d'Houdetot. But
the heart revealings of *Jean Jacques* are its crowning glory.

Just published by

CALVIN BLANCHARD.
76 NASSAU STREET, N. Y.

MOMENTOUS WORK.

THE DOCTRINE OF INSPIRATION, BEING AN INQUIRY CONCERNING THE INFALLIBILITY, INSPIRATION AND AUTHORITY OF HOLY WRIT. By the Rev. JOHN MACNAUGHT, M. A. Oxon, Incumbent of St. Chrysostoms Church, Everton, Liverpool. 12mo. $1.37. Mailed free.

This work is more significant than any which has appeared since the advent of Strauss's Life of Jesus. The vulgar idea of the supernatural inspiration of the Bible is here abandoned; and what is more, it is shown that many of the chief dignitaries, including four Bishops of the Church of England, have held, *on the sly,* similar opinions. The citadel of bigotry, superstition and intolerance, may *now* be considered as *authoritatively* surrendered.

" It is the first book written by an Orthodox clergyman which decidedly denies the doctrine of Scriptural Infallibility. It is well written and manly." *Christian Inquirer.* [*Unitarian*].

From the Westminster Review.

" Distinguished by a fearless investigation of truth, an uncompromising hostility to deception and make-believe. Distinguished likewise by clearness of conception, closeness of argument, purity of expression, and completeness of arrangement. And unless intolerance and superstition shall succeed in smothering the work, it is one which will exercise a wide influence —one which will give form and substance to thoughts which have been floating vaguely in many mens minds —one which will supply a rallying point, and become in lieu of a creed to those who are dissatisfied with traditional and untenable theories respecting inspiration."

Published by **CALVIN BLANCHARD,**
76 Nassau St. New York.

LIBERAL BOOKS

PUBLISHED BY

CALVIN BLANCHARD, 76 Nassau St., N. Y

(SENT BY MAIL POSTAGE FREE.)

COMTE'S POSITIVE PHILOSOPHY, 8vo. pp. 838.......$3 00
COMTE'S SOCIAL PHYSICS.................................... 25
STRAUSS' CRITICAL EXAMINATION OF THE LIFE
   OF JESUS, 2 vols. 8vo.................................. 4 50
FEUERBACH'S ESSENCE OF CHRISTIANITY, 12 mo. 1 50
GREG'S CREED OF CHRISTENDOM, 12mo............... 1 25
HOWITT'S HISTORY OF PRIESTCRAFT, 12mo........ 75
MARY WOLLSTONECRAFT'S RIGHTS OF WOMAN. 75
VOLNEY'S NEW RESEARCHES ON ANCIENT HIS-
   TORY, 12mo... 1 25
VOLNEY'S RUINS, paper cover and bound ....30 and 50
TAYLOR'S PULPIT, 12mo 1 25
TAYLOR'S ASTRO-THEOLOGICAL LECTURES, being
   the second series of The Devil's Pulpit, 12mo........ 1 37
TAYLOR'S BELIEF NOT THE SAFE SIDE............... 10
TAYLOR'S LECTURES ON FREE MASONRY............ 25
WHO IS THE LORD GOD? By TAYLOR............. 30
WHO WAS JESUS CHRIST?............................... 10
WHO IS THE HOLY GHOST? By TAYLOR.......... 10
WHO IS THE DEVIL? By TAYLOR.................... 15
THE NEW CRISIS, or Our Deliverance from Priestly Fraud,
   Political Charlatanry and Popular Despotism...... .. 13
THE ESSENCE OF SCIENCE, or The Catechism of Posi-
   tive Sociology and Physical Mentality. By a Stu-
   dent of Auguste Comte, 12mo.....................60 and 37
HITTELL'S PLEA FOR PANTHEISM................ 25
HITTELL'S PHRENOLOGY................................. 75
WHAT IS TRUTH? or Revelation its Own Nemesis, 12mo. 1 25
MACNAUGHT ON INSPIRATION, 12mo,............... 1 37
VESTIGES OF CIVILIZATION, 12mo.................. 1 25
HITTELL'S EVIDENCES AGAINST CHRISTIANITY,
   2 vols, 12mo................................... 2 50
HELL ON EARTH; or, an Exposé of the Infernal Machina-
   tions and Horrible Atrocities of Whited Sepulcherism :
   together with a Plan for its Final Overthrow......... 18
ROUSSEAU'S CONFESSIONS, Complete, 2 vols, 12
FOURIER'S SOCIAL DESTINY OF MAN, 8vo....$
HOW TO GET A DIVORCE; together with the Law
   all the States in the Union on this subject.......
BOCCACCIO'S DECAMERON, 12mo illustrated......
THE LIBRARY OF LOVE; 24mo. with engravings. T
   most exquisitely amorous and recherche effusio
   ever penned. Comprising :
OVID'S ART OF LOVE, and Amorous Works entire, 50
KISSES OF SECUNDUS AND BONNEFONS, ... 50
DRYDEN'S FABLES...... 50